International Poultry Lib

CALL DUCKS
&
DUCK MANAGEMENT

OTHER TITLES BY JOSEPH BATTY
The Ancona Fowl
Artificial Incubation & Rearing
Aylesbury ducks
Bantams -- A Concise Guide
Bantams & Small Poultry
Brahma & Cochin Poultry
Breeds of Poultry & Their Characteristics
Concise Poultry Colour Guide
Domesticated Ducks & Geese
Hamburgh Poultry Breeds
Indian Runner Ducks
International Poultry Standards
Japanese Long Tailed Fowl
Khaki Campbell Ducks & the Campbells of Uley
The Malay Fowl
Marsh Daisy Fowl
Minorca Fowl
Natural Incubation & Rearing
Natural Poultry Keeping
Old & Rare Breeds of Poultry
Old English Game Bantams
Old English Game Colour Guide
The Orloff Fowl
Orpington Fowl (with Will Burdett)
Ostrich Farming
Practical Poultry Keeping
Polish Poultry Breeds
Poultry Ailments
Poultry Characteristics—Tails
The Poultry Colour Guide
Poultry for Beginners
Poultry Shows & Showing
Races of Domestic Poultry
(With Sir Edward Brown)
Rhode Island Red Fowl
Rosecomb Bantams
Scottish Poultry Breeds
Sebright Bantams
Sicilian Poultry Breeds
The Silkie Fowl
Sussex & Dorking Fowl
True Bantams
Understanding Modern Game
(with James Bleazard)
Understanding Indian Game
(with Ken Hawkey)

CALL DUCKS & DUCK MANAGEMENT

Dr Joseph Batty

Beech Publishing House
Station Yard
Elsted Marsh
MIDHURST
West Sussex GU29 0JT

© Joseph Batty, 2005

This book is copyright and may not be reproduced or copied in any way without the express permission of the publishers in writing.

ISBN 1-85736-427-9

First Edition 2005

British Library Cataloguing-in-Publication Data
A catalogue record for this book is available from the British Library.

Beech Publishing House
Station Yard
Elsted Marsh
MIDHURST
West Sussex GU29 0JT

CONTENTS

1. History	7
2. Calls Described	21
3. Standard Colours	31
4. Breeding	43

Duck Management

5. Breeds of Duck	49
6. Accommodation	59
7. Feeding	75
8. Incubation & Rearing	91
9. Management Pointers	103
INDEX	117

Colour Plates   Throughout (3)

FOREWORD

Ducks are fascinating to keep and provide a profitable business for those who have the land and buildings, or can be kept on a hobby basis to provide eggs and table birds.

Despite what is thought essential, the provision of a pond, this is not vital and, in any case, small fibreglass ponds can be purchased and installed quite easily.

If the duck-keeper wishes to exhibit birds there is usually provision for these at the major shows. They can be bred for fancy show points and win prizes which allows them to be sold at high prices.

Call ducks are now very popular and available. From the days when they were simply Decoy Ducks for use in catching wild waterfowl, they have been developed into very attractive, small ducks, which can be kept by anyone with a garden or orchard.

My thanks to Paul Chapman, Poultry Artist, who painted the Call Ducks so the wonderful colours now available can be seen.

Joseph Batty   Elsted Marsh
                                October, 2004

I

HISTORY OF
THE CALL DUCKS

Call Ducks Around 100 Years Ago
At first they were in two varieties -- Greys and Whites. Then came many new colours which are now well established.

EARLY HISTORY

The Call Duck has a long history, but is best known for two features:

1. Sportsman's Decoy Duck.
2. Ornamental domesticated duck.

These are considered below. They are believed to originate in Holland, although they are quite widespread so to designate a specific area of origin may be misleading. In fact, there are references to the early *breeding* of call ducks in Japan and China. However, the reference to the origin may be *to the use of the Call duck for sporting purposes* because it appears the Dutch invented the procedures followed.++

Recognition as a standard breed occurred in the USA in 1874 for the Grey and White and the other colours much later.

Lewis Wright believed they were small descendants of the Mallard, and, at the time, found they were kept very little, except in public parks.* How situations change because now they are quite popular, hence the proliferation of varieties!

Another authority**, states there are the White and the Grey Call Duck. They are, in effect, bantams, being miniatures of a larger breed. As their name implies, they are "remarkable for their loud and continuous quacking, in a shrill high note, which can be heard at a great distance, and which renders them admirable as decoy ducks to allure the wild species to their destruction."++ He notes they are also very attractive as ornamental waterfowl.

The Book of Poultry, Rev. Ed. S H Lewer, c. 1920
** The Poultry Book, W B Tegetmeier, 1873.
++ *Our Poultry*, Harrison Weir, 1903.

CALL DUCKS

Another name given to the breed is the "Dutch Dwarf Duck"* thus acknowledging that they were possibly first used in Holland as decoys. In fact, this is acknowledged by the famous author Francis Willughby (1676) who records that pools are used, fitted with channels and nets, in which were placed the Decoy (Call) ducks. As a result, Mallard, Teal, Wigeon, and other wild ducks were enticed to the scene and captured in the nets.

The Call ducks are placed outside the cylindrical nets, but by their quacking entice the wild ducks to fly in and take the bait which is placed for them. The procedures differ, sometimes the Call ducks are pinioned or one side of the wing feathers cut so they cannot stray; at other times they are left intact and, since they are quite domesticated, do not fly away. They may also fly to other ponds and return with other species of duck, thus allowing them to be captured. They are, in effect, trained to act as decoys.

Again the reference is made to their being developed from the Mallard, but no explanation is made on how the breed developed the persistent, loud quacking which allows them to attract other ducks. Possibly, over a long period, the loudest quackers (ducks) were selected, thus developing the feature. The Mallard on a pond is quiet.

Another writer** states that "this is the bantam of its race, usually coloured like the wild mallard, but often *white* -- a colour preferred by fowlers, who use it in the decoys, on account of its being easily distinguished from the others. These

---

* Harrison Weir, (ibid)
** John Sherer, *Rural Life,* c. 1800. Reference to 'Crests' appears to mean *breast.*

Japanese Call Ducks

birds are a Wiltshire breed, with compact and elegantly rounded *crests (?)* , and are very handsome."

The reference to Wiltshire is misleading because the breed is very much a part of rural life in many counties in this country and throughout the world. Another source gives examples from Lincolnshire and supplies drawings of keepers at work. These are reproduced below.* A description of the process of decoying is repeated from the same source.

Decoying

In the lake to which the wild ducks resort, their most favourite haunts are observed. Then in the most sequestered part of this haunt a ditch is cut, which is about four yards across at the entrance, and decreases gradually in width from the entrance to the farther end, which is not more than two feet wide.

The ditch is of a circular form, but does not bend much for the first ten yards. The banks of the lake on each side of this ditch (or "pipe," as it is called) are kept clear from reeds, coarse herbage, &c., in order that the fowl may get on them to sit and dress themselves. Along the ditch, poles are driven into the ground, close to its edge, on each side, and the tops are bent over across the ditch and tied together.

These poles, thus bent, form at the entrance of the ditch or pipe an arch, the top of which is ten feet distant from the surface of the water. This arch is made to decrease in height as the pipe decreases in width, so that the remote end is not more than

---

* *Museum of Animated Nature*, Charles Knight, c. 1800

eighteen inches in height. The poles are placed about six feet from each other, and connected by poles laid lengthwise across the arch and tied together. Over the whole is thrown a net, which is made fast to a reed-fence at the entrance and nine or ten yards up the ditch, and afterwards strongly pegged to the ground. At the end of the pipe farthest from the entrance is fixed a "tunnel-net," as it is called, about four yards in length, of a round form, and kept open by a number of hoops about eighteen inches in diameter, placed at a small distance from each other to keep it distended.

Supposing the circular bend of the pipe to be to the right when one stands with his back to the lake, then on the left-hand side a number of reed-fences are constructed, called "shootings," for the purpose of screening the "decoy-man" from observation, and in such a manner, that the fowl in the decoy may not be alarmed while he is driving those that are in the pipe. These shootings, which are ten in number, are about four yards in length, and about six feet high. From the end of the last shooting a person cannot see the lake, owing to the bend of the pipes, and there is then no further occasion for shelter. Were it not for these shootings, the fowl that remain about the mouth of the pipe would be alarmed if the person driving the fowl already under the net should be exposed, and would become so shy as entirely to forsake the place.

The first thing that the decoy-man does when he approaches

Decoying in Operation
This was taking place in the Fens in Lincolnshire

the pipe is to take a piece of lighted turf, or peat, and hold it near his mouth, to prevent the birds from smelling him.

He is attended by a dog, trained for the purpose of rendering him assistance. He walks very silently about half-way up the shooting's, where a small piece of wood is thrust, through the reed-fence, which makes an aperture just large enough to enable him to see if any fowl are in ; if not, he walks forward to see if any are about the entrance of the pipe. If there are, he stops and makes a motion to his dog, and gives him a piece of cheese, or something else, to eat; and, having received this, the animal goes directly to a hole through the reed-fence, and the birds immediately fly off the bank into the water.

The dog returns along the bank between the reed-fences, and comes out to his master at another hole. The man then gives him something to reward and encourage him, and the animal repeats his rounds until the birds are attracted by his motions, and follow him into the mouth of the pipe.

This operation is called "working" them. The man now retreats farther back, working the dog at different holes until the ducks are sufficiently under the net. He then commands his dog to lie down behind the fence, and going himself forward to the end of the pipe next the lake, he takes off his hat and gives it a wave between the shootings.

All the birds that are under the net can then see him; but none that are in the lake can. The former fly forward, and the man then runs to the next shooting and waves his hat, and so on, driving them along until they come to the tunnel-net, into which they creep.

When they are all in, the man gives the net a twist, so as to prevent them from getting back. He then takes the net off from the end of the pipe, and taking out, one by one, the ducks that are in it, dislocates their necks. This is the scene represented in the drawing. The net is afterwards hung on again for the repetition of the process; and in this manner five or six dozen have sometimes been taken at one drift.

When the wind blows directly in or out of the pipes, the fowl seldom work well, especially when it blows into the pipe. The reason for this is, that the ducks always prefer swimming against the wind, otherwise the wind blowing from behind catches and ruffles their feathers. If many pipes are made in the same lake, they are constructed as to suit different winds, and are worked accordingly. The better to entice the fowl into the pipe, hemp seed is occasionally strewn on the water.

Mallard -- a close relation to Call ducks

CALL DUCKS

Duck Shooting

The Decoy duck can also be employed to attract wild birds for shooting. The constant quacking of the females - the "call" - brings in the Mallard who, unsuspecting, fly in to join other ducks who are feeding.

This has been summed up as follows:

> "Exceedingly good sport may be obtained, especially on a very stormy day, by placing a few decoy ducks within good range of a gun stationed near the bank of a sheltered bay or eddy. The decoy can also be used with success when ducks are flighting to crop or roots."*

Decoying can be used in different forms from the hunter mimicking the call of ducks, to using a specially made Whistle, or by the means of Decoy ducks here described. Artificial duck decoys are also available in wood or other material which floats, thus attracting wild ducks down.

Wild ducks will not approach if there are signs of keepers or others with guns. Therefore the trappers must be hidden away, concealed in bushes or in a specially constructed Hide, which fits in with the landscape.

Wildfowlers build huts made of branches, one for each person. He or she sits there and when birds approach, will imitate the call or "cackle", thus luring them onto the pond or lake.** As they approach and he can see their eyes he fires his gun. Artificial decoys may also be used to attract the wild fowl.

---

* *The Keeper's Book,* Sir Peter Jeffrey Mackie, Bart.
** John Sherer, *Rural Life,* (ibid)

Taking A Shot

At the Ready

LINKS WITH THE MALLARD

As noted, there are many links with the Mallard *(Anas boschas)* from which the domesticated variety were developed. Certainly, there are similarities.

Mallard
    Size: Drake or Mallard:
    Length 23 inches Weight 36 to 40 ounces (1020 - 1133g.) the drake being larger.

Call Duck
    Size: Call Ducks
    20 oz to 26 oz (570   740g.) for duck and drake respectively.

For both breeds the young will obviously be lighter in weight. The full weights are usually reached at 12 to 18 months depending upon conditions and food available.

From these figures it will be seen that the Call Ducks are much smaller. Taking the drakes the figures are 26 to 40 oz ; ie, 14 oz smaller or about two-thirds the size of the Mallard.

The description of the Call Duck and its many varieties is covered in the next chapter. They are very similar to the Mallard, although the Calls are much smaller and are rounder, a shape known as "cobby". This occurs with many breeds of poultry and in some cases the breadth is almost as much as the length. Old English Game bantams* are probably the best example which has been developed from a fairly long bird to one which is triangular-shaped across the back.

* *See Old English Game Bantams*, Joseph Batty.

CALL DUCKS

This shortness of back is the result of careful, selective breeding over a period of about 100 years. In the process the normal sized tail has been lost. However, It is obviously very important, with any breed of bird, not to carry out modifications too far; there should be harmonious balance or beauty and usefulness are lost.

DESCRIPTION OF THE MALLARD*

MALE—Head and neck brilliant-metallic green, the lower part encircled with a pure white ring; back and upper breast rich chestnut; remainder of breast and under parts greyish white, finely pencilled with dark lines; four centre feathers of the tail stiffly curled upwards; wing-bar blue, violet, and green-- feet orange; bill yellowish green.

FEMALE-- Plumage of various shades of grey and brown-white throat; without the curl in tail.

YOUNG—In down, olive yellow, with a stripe on the forehead, passing over the eyes to the back of the head.

EGG—White, washed with green; eleven to sixteen- per clutch.

BREEDING -- March—May. Incubation, twenty-eight days.

Nest Down -- Neutral grey in colour, occasionally with a slight tinge of brown.

---

* *Ornamental Waterfowl*, Hon. Rose Hubbard, London

2

CALL DUCKS
DESCRIPTIONS
&
STANDARDS

CALL DUCKS

STANDARDS

The *Standards* are written descriptions of each breed of poultry, carefully worded, so they can be understood by fanciers and poultry judges. In the English language the American and British standards were drafted at the same time (around 1870) although there are now variations to be found between them.

Over time, new breeds and varieties (sub-divisions of breeds) have been added. In the case of the Call Duck the new varieties have been considerable. At first there were Greys and Whites. but now there are many varieties, possibly around fifteen different colours, as well as 'off-colours' which are colours obtained when breeding two colours together, but finish up with a non-standard colour.

Many of the varieties are not fully stabilized so variations in colours will occur. If not exactly correct the judging should be on the basis of 'Any Other Colour' if that is permitted by the particular show regulations.

CALL DUCKS

DESCRIPTIONS

Call Ducks have been described in many ways; thus;
1. **Bantam Ducks**
2. **Dwarf Ducks**
3. **Small, Cobby Ducks or Miniatures.**

They must be quite small and plump. One author* sums up the requirements:

> They might indeed be termed Wild Duck Bantams. *They have short little bills and prominent foreheads, their heads having a round appearance. The aim of the breeder is to get them as small as possible, tight in feather, and good in shape, colour and condition.*
>
> They are continually "quacking" in their shrill note, which can be heard for a considerable distance. It is this characteristic which makes them so valuable as " call" ducks, for they are much used in wild duck shooting, their continual and loud quacking attracting the wild birds. They are also called " coy " or " decoy " ducks, as they are the means used for enticing the wild ducks into the decoys the wire enclosures or traps for catching them. They are hardy and do well in most localities, but the newly-hatched ducklings are very small and need special care. They should be kept in a dry place for the first week or so. Although small for the purpose they are excellent as table birds, for the flesh has a good flavour and is very sweet and delicate.

Since it has evolved from other breeds, most probably the Mallard, it breeds freely with other domesticated breeds and, if purity is to be achieved, they should be kept separately from the others. Otherwise, the breeding to type will be lost and mongrels will emerge.

---

* *Ducks & Geese*, F J S Chatterton, London

CALL DUCKS

DESCRIPTIONS

The various parts of the Call Duck are now explained with notes on the desirable features and possible faults. In the next chapter the colour requirements are explained.

BODY

This is set an a moderate angle to the tail and is shaped like a cone, which is well rounded at the front and comes to a graceful point (not too sharp) at the tail, which is short and follows the back at the top and the upward curve of the abdomen on the underside. It is sometimes suggested that the carriage should be horizontal, but this is unrealistic because these ducks do have a slight uplift at the front.

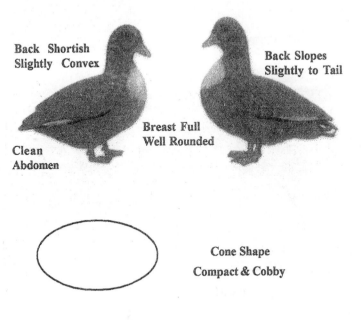

Wings & Tail

The wings are quite full with long main feathers because they do fly. They are carried high so the top edge rests along the top side of the back, but should not be above the back or crossed above the tail. If too high this is known as "Goose Winged" and is a fault.

Neither should there be feathers out of alignment on the wing, or the wing to have a gap in the middle of the wings (split wing).

Some judges of ducks do not handle the birds -- as usually is the case with fowls -- on the grounds that condition can be seen by viewing in a pen. Whether this is wise is questionable because if wing feathers are broken, the bird should be penalized; in any event only handling will reveal condition.

Faults on Call Ducks

CALL DUCKS

HEAD & NECK

Head should be round with a wide skull, and pronounced cheeks. The roundness should be obvious and not just an indication of a circular head and face.

This feature differs from all other species of domesticated duck. The Mallard is relatively long and so are the others. The Rouen tends to be rounder, but not as compact as the Call Duck. Strangely, the head of the Carolina Duck is somewhat similar, but nobody has suggested that this breed was ever involved in developing Call ducks or whether this would even be possible.

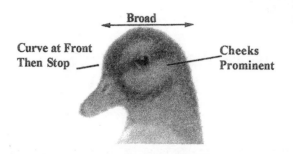

Head of Call Duck

The earlier types did not have such round heads, and, obviously, these have come about from selection to meet show requirements. Those which are regarded as utility or pet birds, with no inclination to show, will probably have longer heads, along the lines of the Mallard.

The example shown opposite from a Japanese Call Duck illustrates where the head has not been refined. As noted, heads do differ considerably, even from one variety to another. There is a tendency for the Grey variety to be larger than, say, the White, both in body and other features.

CALL DUCKS

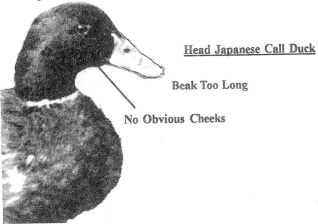

Long Head, Lacking Forehead

Head Japanese Call Duck

Beak Too Long

No Obvious Cheeks

Head of Utility Japanese Call Drake.

There are many possible shapes to be found:
1. Squarish 2. Pointed 3. Narrow 4. Narrow, etc

The acceptable type of head can be shown in the form of circles or diagrams:

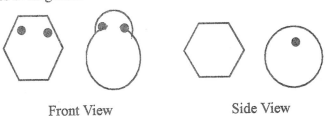

Front View   Side View

The six-sided shape needs to be rounded at the sharp corners, but it does give the general idea on shape. Any shape which depart from the globular shape is incorrect and should be penalized when judged.

CALL DUCKS

The Neck

The neck should be quite short and fairly thick set, with a slight arch.

Long necks or those with a curve are not normal for a short, cobby bird like the Call Duck.

The Bill

The bill should be fairly short and wide for its size, with a slight concave slope along the top. However, this downward dip should be very slight or the bill will be termed 'dished' which is a fault. The British Poultry Standards suggest not longer than 3.10 cm (1.25 inches), but do not specify what is the minimum length, which is important -- a duck must have an adequate bill for collecting food and other purposes.

Disfiguring marks or scars or the wrong colours are also faults. The colour should match the plumage colour and this is covered in the next chapter dealing with the colour standards.

The Eyes

Eyes should be relatively large and bright and set midway to moderately high in the head; some suggest fairly low, but this makes birds look rather sneaky. Usually the eye is set slightly to the front of the face on a line which denotes the top of the cheek, which is in line with the top of the bill.

Watch out for any encrusting around the eyes which will tend to occur when the ducks do not have clean water in which to dip their heads.

Bill colours vary from a yellow to orange to black and should be appropriate to the specific variety, any departure being a fault. What the colours should be is covered in the next chapter which deals with colour standards.

CALL DUCKS

Carriage

The emphasis is usually on achieving as horizontal carriage as possible, but not more than 25° above horizontal is a more realistic approach.

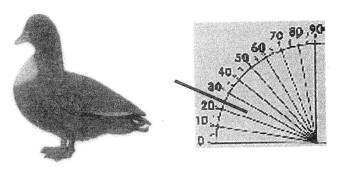

This drake is around the maximum above horizontal.

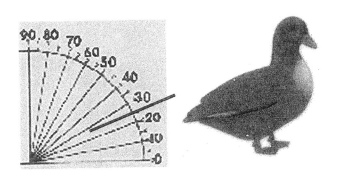

This duck stands fairly upright and, any more, should count as a fault.

CALL DUCKS

LEGS

The legs should be short and positioned centrally with the feet fully webbed. The front toes should be straight with nails intact. The colour of the shanks varies from light orange to dark, with the dark varieties having darkish legs. They should be smooth without discolouring marks or blemishes.

As noted, the legs should be of short, with muscular thighs. If the legs are too long the keel will be raised and will not be top exhibition stock, but may be more fertile. Toes should be webbed and moderately large, set firmly on the ground to support the body weight. There should be no deformities.

Long Shank   **Short, Thick Shank**

The long shank will be found on tall ducks which will certainly be larger than the Call Duck. For the latter the short, thick shank will be appropriate, but this does not indicate that the breed is slothful because the breed is quite active and, as a result, full of character and interest.

CALL DUCKS

CRESTED CALL DUCKS

Crested Ducks are a separate breed, but, no doubt due to crossing with a Call Duck the crest has been introduced. They have been in existence for a considerable period.

The history of the normal Crested Duck is well recorded. The following is relevant:*

> The Crested is a medium sized duck (around 6-7 lb - 3 k.). In attempting to describe the type the British Standard states that the body and characteristics of the Crested are rather like the Orpington. If the crest is to be perpetuated and improved upon, it is essential to breed from birds with large, globular crests. Usually some ducklings will be hatched without crests. Although these birds, when mature, will breed crested progeny, they are better discarded as breeders. Indeed, for breeding, the largest crests should be present, although some breeders will state that the presence of a large crest on *both sides* is unnecessary.
>
> Apparently the crested variety has lethal genes which can result in approximately 25 per cent dead in shell. This means that a larger number of eggs must be set to obtain the number of ducklings required.
>
> Care should be taken when rearing the ducklings with large crests. Some breeders suggest that the youngsters may be frightened by their crests and rush around, thus causing injury to themselves. If this happens, the fluff may be clipped from the scalp.

Whether the lethal gene would affect the Call Duck is not

---
* *Domesticated Ducks & Geese*, Joseph Batty, BPH

clear, but it is possible that a cross between a normal Crested and a Call Duck, being crested in a diluted form, would not introduce a strong lethal gene. However, if *Crest X Crest* took place it may be a different story. An example of a full crest is given below:

Head of normal Crested Duck

In the Call Duck the crest would be as shown above, but the head would be different. The requirements are covered earlier in the chapter.

Various claims have been made for developing the Crested Call Duck. Deserving special mention is Dr C Darrel Sheraw, a leading authority in the USA, who took 20 years to achieve excellent results. He has managed to get large, round crests, which are still compatible with the cobby body and globular head of the Call Duck. This is a great achievement because all features should be well balanced when presenting a duck for judging !

---

* *The Call Duck Breed Book*, Augusta, 2003

CALL DUCKS

JUDGING CALL DUCKS*

When judging Call ducks it is necessary to look for type. Accordingly, attention should be paid to the main features, establishing to what extent actual and standard requirements agree. More specifically the following should be examined:

1. Size

Should be a small cobby bird; and any feature which departs from the norm should be penalised; for example:

FAULTS:
- (a) slim bodied;
- (b) tall;
- (c) wings long and loose;
- (d) lack of full breast;
- (e) carriage not nearly horizontal;
- (f) long back;
- (g) roach or indented back;
- (h) tail lacking neatness;
- (i) faulty feathering, including too fluffy;
- (j) faulty wings.

2. Head and Neck

The head is required to be round in shape and set on an extremely short neck, with pronounced cheeks. Accordingly, a long neck or a small head would be unacceptable. A bill which is narrow or long would also constitute a major defect. Eyes in the wrong position or defective would be regarded as a fault.

---

* Batty *(ibid)*

DISQUALIFICATION OR BEING 'PASSED'

Whether a bird should be disqualified for a specific fault is difficult to give a positive answer in all cases. Simply being of poor quality is insufficient a reason for that drastic step.

Evidence of cheating such as the removal of feathers, the use of dyes, absolutely the wrong colour in a specific class (this may be passed rather than disqualified), major faults, signs of disease or infection or wounds, and any other serious condition or departure from the standard may call for disqualification.

There should be agreement by at least two judges if a bird is to be disqualified. Otherwise there could be serious disagreement.

The *American Standards of Perfection* lay down rules for disqualification, but some of these, such as faulty colour, appear rather harsh. Obviously, faults should be penalised, but the many of the colours tend to vary and be unstable, so disqualification is rather strong. In fact, there is no universal acceptance of some colours and new ones are constantly appearing. In Britain *Colour* carries 20 points and, provided a bird is near enough the correct colour, all that should be done is for points to be deducted. However, if the colour is quite wrong, the bird should be 'passed'.

3

STANDARDS COLOURS & VARIETIES

CALL DUCKS

NOTE ON BREEDS & VARIETIES

A *breed* is a type of duck with specific characteristics and usually recognized by poultry clubs or societies. It breeds with the same features on a consistent basis, so the breeders know what to expect.

Put another way it is a group of birds which have an identical gene structure.

Within a breed there may be variations, usually in the colour or markings. These are known as *"Varieties"* being sub-species of the main or original type which, in Call Ducks, were the Grey (Mallard type) or the White.

Cross Breeds

If different breeds are allowed to mate together these become cross breeds and would not be allowed to be judged at a show unless in a non-standard class. This practice is not to be recommended because the 'pure' breeds would be lost if carried too far. In the wild the separate species do not generally inter-breed.

The domesticated ducks can cross because they come from a single source -- the Mallard. However, occasionally, different claims are made that Call Ducks may have been developed from a different water fowl, but this has not been substantiated.

SPECIAL NOTE ON COLOURS

Many of the colours are difficult to describe with great accuracy, the coloured birds (other than White) are a variation of the Mallard/Wild duck. Reference should be made to the standards or to the Call Duck Association, which is the breed club. Originally, there appear to have been two colours, the Grey and the White (*The Poultry Book*, W B Tegetmeier, 1873) so the other colours have been Sports or crosses.

COLOUR STANDARDS

Any departure from the wild duck colour, such as no white neck ring or absence of deep claret colour in breast (in the drake), should be penalised. With the self colours there should be no other colour, other than White, Blue or Buff as the case may be.

RECOGNIZED COLOURS

As noted, colours are now quite numerous and more keep being developed.

1. Apricot
2. Bibbed
3. Black
4. Blue
5. Blue Fawn
6. Buff
7. Grey (Mallard)
8. Magpie
9. Pastel
10. Pied
11. Silver
12. Snowy
13. White
14. Non-standard.

Some appear in the USA Standards and others in the British Standards. Examples of colours which are not fully standardized are: Butterscotch; Spotted, Yellow-bellied, Cinnamon, Harlequin, and others. Unfortunately, the process appears to have gone too far with resulting confusion and lack of cohesion and co-operation between the various clubs and societies. This state of affairs discourages fanciers who find the descriptions and their variations quite bewildering.

BRIEF DESCRIPTIONS OF COLOURS

APRICOT

Duck is an Apricot colour with light grey and pearl grey on wing feathers. The drake is largely a grey colour, dark above and light below. The head is silver-grey and the breast (large bib) a sort of mulberry colour. Bill light green in drake and light brown in duck. Orange shanks and feet.

FAULTS: White in wings; wrong colour bill or shanks.

BIBBED

This breed take their name from the fact that they have white bibs with a main colour of black, blue or lavender. They are similar to the Swedish or Orpington ducks which are bibbed, although, obviously, not with the same breed characteristics. Drake bill is orange and duck black. Legs orange, but shaded with dark.

FAULTS: Poor bib; wrong colour bill or shanks.

BLACK

This variety has been developed and should be entirely black with dark legs and bill. No other colours should appear in the plumage.

FAULTS: White in wings; wrong colour bill or shanks.

CALL DUCKS

BLUE

Similar to the Blue Swedish Duck in colour. Legs orange black; bill light brown. May be bibbed as the Swedish, but there seems no reason why an overall Blue cannot be bred. Really a Blue Bibbed, although may be a different shade.
FAULTS: White in wings; wrong colour bill or shanks.

BLUE FAWN

Duck is brown, rather similar to the wild duck and the drake is marked like the Mallard but the colours are different -- head dark blue (instead of green) and body grey, dark at the top and silvery grey below. Bill light brown and legs orange.
FAULTS: White in wings; wrong colour bill or shanks; broad lacing on duck's back; other colour defects.

BUFF

A nice even buff or seal brown all over. Bill yellow; eyes brown.
FAULTS: White in wings; wrong colour bill or shanks.

GREY (MALLARD)

The duck is similar to the wild duck and the drake follows the Mallard except the upper body is grey and the lower part light silvery grey. The drake has the blue wing bar. Legs are orange with a darker tinge. Eyes brown.
FAULTS: White in wings; wrong colour bill or shanks; white ring on duck; no white ring on drake.

CALL DUCKS

MAGPIE

The colour pattern is similar to the large Magpie breed of large ducks with black or blue crown and on wings, same on back and tail. Remainder white. Legs orange with dark tinge.

FAULTS: Lack of positive areas of colour; foul feathering; wrong colour bill or shanks.

PASTEL

The colours are quite beautiful. The duck is a yellowy-cream colour with lighter markings across the back and a broad wing bar. The drake follows the Mallard pattern but with powder blue for the head, brown for the breast, upper back is creamy fawn and the lower part a light colour with faint lacing. The wings are light grey with a blue bar. Shanks orangy-salmon and eyes brown. Bill on drake yellowy-green and female darkish brown.

FAULTS: White in wings; foul feathering; wrong colour bill or shanks.

PIED

Similar to Mallard, but with even markings/ splashes making the plumage pied. Drake's bill yellow with green overlay; duck's bill is yellow.

FAULTS: Wrong colours ; wrong colour bill or shanks.

SILVER

Follows the Mallard, but breast feathers on drake have white lacing. Back lightish grey. Bill is green.

Duck is a fawn and coloured, with markings, such as mottles, in brown and grey. Bill on female is orange/brown.

FAULTS: Foul feathering; wrong colour bill or shanks; pencilling on duck; irregular markings on head of drake.

White Call Ducks in Natural Surroundings
These small, gentle ducks were painted specially for the publishers to indicate the many colours now available.

Call Ducks
(Painted by Paul Chapman)

Call Ducks
(Painted by Paul Chapman)

Call Ducks
(Painted by Paul Chapman)
Colours do vary to some extent.

CALL DUCKS

SNOWY

This breed has a mixture of colours. The duck has a fawny-brown head and a whitish body which is marked with fawn marbling on the breast and indistinct grey triangles on the back.

The drake has a deep green head with markings like the duck on the whitish body, but much deeper in colour; in fact, the breast markings are half-moon shape in a dark fawn colour.

Drake's bill yellow with a green tinge; female bill dark orange. Eyes: brown. Legs orange with darker tinge.

FAULTS: Foul feathering; wrong colour bill or shanks.

WHITE

This is a purely white breed. Shanks bright orange. Eyes blue. Bill bright yellow. In effect, both sexes the same except for drakes curled tail feathers and possibly size.

FAULTS: Foul feathering; wrong colour bill or shanks.

OTHER COLOUR FEATURES

Colours on bill and feet vary according to the colour of the plumage. Some of the coloured drakes -- those similar to the Mallard --have a white ring on the neck.

These colours will not necessarily breed true to type. The Grey and White should breed true, but with mixed colours; eg, Mallard to Blue Fawn the percentage will tend to be 50 : 50. Blues in any poultry are difficult to breed true.

CALL DUCKS

At the early stages of a breeding programme many strange colours may appear and only when the breeding has continued for a few years will it be possible to anticipate results with reasonable certainty. Some inbreeding may be necessary to lock in the desirable features and, at all times, great patience will be required. A desired colour may not be present at the first breeding, but appear in the second or third time round.

NOTE: There will be variations when drakes are in eclipse plumage.

FAULTS: The possible faults listed indicate what foul feathering may result in penalties, maximum 20 points. Wrong colouring in the bill, shanks, feet, or eyes may also call for penalties.

4
NOTES ON BREEDING CALL DUCKS

Tranquility

Ducks on a large pond. This would be found on a large estate, but a simple small pond is adequate for keeping a few Call ducks.

CALL DUCKS

BREEDING

Call Ducks are not prolific layers and, therefore a good yield is probably around 60 eggs per season. This is not many compared with, say, Khaki Campbells or Runner Ducks, but they are only bantams, so they follow the same pattern as other bantams.

In any case they will eat much less food than their commercial cousins and still become plump enough to make small table birds.

Mating Up

Some breeders run a single drake with three or four ducks and this usually works provided the birds are in good condition. An alternative is to create competition and let two drakes run with a pen of eight to ten ducks.

The birds are placed in breeding pens early in the year and may start laying shortly after. If the ducks lay very early on -- even before Christmas -- a few eggs can be tested for fertility by popping in an incubator for 7 - 10 days and then candled (tested by holding egg before a bright light) and provided a reasonable number are fertile, they may be incubated.

However, do not be too optimistic early on because birds are not ready for breeding until the weather is beginning to get warmer. Accordingly, the eggs may not be fertile, or, if they hatch, the ducklings may be weak and, not withstand the colder weather, especially if there is frost around.

Eggs should be collected daily and placed in an incubatator, duly marked with date and pen number. Try to get enough eggs together within 7 - 10 days to make it economical to run an incubator and mark with a cross on one side to facilitate turning. Then candle at 7 days and, again, at 14 days, removing any without embryos or obviously dead embryos. Then get ready for the remainder to hatch at 28 days.

A broody hen may be used to hatch the eggs and she will have the advantage of being able to rear the ducklings whereas with an incubator some form of rearer will be required. However, finding a broody early in the season may be difficult, although a Silkie or Silkie cross tends to come broody throughout the year.

Some breeders allow Call duck females to hatch their own eggs, but this will certainly reduce the egg output and the other methods may be more economic.

ACCOMMODATION

Only a small enclosure will be necessary to keep a pen of Call Ducks and, although a pond is not essential, much enjoyment will be lost by not having one. Seeing them going in and out of the water and swimming around gives great pleasure.

The section on general management covers the necessary procedures, but remember a bantam duck is very small and therefore can be carried away quite easily. The ducklings are also prey for jackdaws and magpies who will also take the eggs.

For larger predators proper fencing will be advisable with precautions taken to ensure that foxes or dogs cannot get into the pens. In the belief that Call Ducks will not stray far, it is tempting to have a low fence, but this does not keep out predators, so great care should be taken to ensure maximum safety. Rats can be a great nuisance and they will chew wood and burrow underneath to carry off ducklings and eat the food that is left around. Digging a trench with wire netting and stones at the bottom should ensure that access cannot be gained. See page 65.

Infra-red Rearing of Ducklings

CALL DUCKS

FEEDING

Feeding is covered later in the Management Section. Here it is important to note that a high protein food is essential:

1. Feeding Ducklings

At the early stages feed turkey or chick crumbs which contain a very level of protein which brings on the ducklings at a fast rate. Quick growth at the beginning is vital because they can be weaned on the normal foods once they are quite plump and active. Chopped boiled egg and finely ground oats with milk may also be given.

Like all bantam poultry great care is needed for the first week or so. Otherwise, they can contract disease and possibly die. But once they are developing they will mature very quickly, within a few weeks. They are best kept in a small rearing shed or, if with a broody hen, in a coop with run attached, but the top covered over. If on short grass the run can be moved around so fresh grass is available.

2. Laying Birds

In the laying season a special ration will be advisable. This can take the form of small size Layers' Pellets supplied in a gravity hopper and fed *ad lib*. Corn and greenstuff should also be provided thus giving a balanced diet.

Water

For the ducklings, water may be supplied in a saucer and then in a chick fount. It is important to ensure in the early stages that the water is not too deep or the ducklings may drown.

Later the larger water fountain can be provided and, when mature, a small pond will be useful and attractive. The ducks love to swim around and derive great enjoyment in the process.

5
BREEDS OF DUCK

Drawing of Call Ducks
These are ornamental ducks

Breeds Compared

Duck Management

Introduction

The main breeds are listed and described in alphabetical order. Certain breeds are kept in considerable numbers whereas others are quite scarce.

List of breeds

Abacot Ranger
Aylesbury
Baldwin
Bali
Black East Indian
Blue Swedish
Blue Termonde
Call
Campbell
Cayuga
Coaley Fawn
Crested
Indian Runners
Huttegem
Khaki Campbells
Magpie
Muscovy
Orpington

Originated in what was Flanders
A fine, hardy table breed

Pekin
Rouen (and variations)
Saxony
Silver Appleyard (Standard)
Silver Appleyard (Bantam)
Stanbridge White
Welsh Harlequin
White Campbells
White Pennine

SILVER APPLEYARDS

BREEDS IN POPULAR DEMAND

This list represents the main breeds which have been kept, but not all are in popular demand, and some are now so scarce that there may be difficulty in obtaining them.

The most popular are:

1. Aylesbury—a table duck.

2. Campbell, mainly Khaki Campbells—these are the best layers, some achieving over 300 eggs.

3. Indian Runners, which are again layers, with little or no carcase for the table.

4. Orpington, although scarce -- a dual purpose breed.

5. Pekin a fairly upright duck which is primarily a table duck.

6. Rouen which is a slow maturing table bird.

The remainder may be found in different parts of the country, especially at the larger poultry shows, but they have to be found. Some are very attractive and are worth the search.

Duck Management 53

Aylesbury Ducks -- mature quickly

Rouen Ducks -- excellent table birds but slower to fatten.

NATURE OF DUCKS

Ducks are hardy creatures and can thrive in conditions that would deter normal breeds of fowl such as Leghorns or Orpingtons. Moreover, many of them lay extremely well, although some people never seem to take to duck eggs, even though they are quite nutritious and delicious.

The fact remains that given the market or need ducks can be quite profitable. Moreover, the heavier breeds, such as the Aylesbury, Rouen, Orpington or Pekin make excellent table birds and these are in constant demand in a gourmet type market.

I. CLASSIFICATION

Attempts have been made to classify ducks into different categories as follows:

LIGHT (5lb; 2.25k)

Abacot Ranger

Bali

Khaki Campbell (almost Medium)

Coaley Fawn

Magpie (almost Medium)

Runner

Duck Management

American Pekin Ducks -- excellent table birds

White Muscovy Ducks -- unusual breed which is quite heavy

MEDIUM (6lb; 2.70k)

Orpingtons

Blue Swedish

Cayuga

Crested

Huttegem

Orpington

Penguin

Saxony

Stanbridge White

Welsh Harlequin

Head of Crested
Rather similar to
Orpington in stature.

HEAVY (8lb +; 3.60k+)

Aylesbury

Baldwin

Blue Termonde

Pekin

Muscovy

Rouen

Rouen Clare

Saxony

Silver Appleyard (Standard size)

**Drawing of
Pekin Drake**

Duck Management

BANTAM SIZE (40oz; approx 1k)
Call or Decoy
East Indian (recognized by APA)
Mallard
Silver Appleyard Bantam Duck

KEEPING THE BREEDS PURE

The haphazard breeding of livestock is not recommended for the simple reason that there is no standard at which to aim. This means "standard" in the broad sense to include the number of eggs, size of carcase and other physical requirements.

Left to their own devices ducks will revert to an almost wild state so the carefully developed attributes which have taken many generations will be lost.

Keeping a specific breed ensures that the produce remains the same or very similar. If a specific size is required for the table within a definite period then the Aylesbury, Pekin or Orpington would be appropriate. If wanting utility, so that eggs as well as a good carcase is obtained then Orpingtons might be the answer. Rouens might produce a top class table bird but it takes longer to mature and therefore is more suitable for the gourmet market.

For Layers a choice would probably be made from Khaki

Campbells or Runner Ducks.

If for showing then one of the more unusual breeds might be selected such as the Cayuga, Saxony or Welsh Harlequin. The selection must be left to the individual and his or her idea of beauty.

Indian Runner Ducks
These are utility type not the bottle-shaped exhibition duck.

6

ACCOMMODATION

East India Ducks & Drake

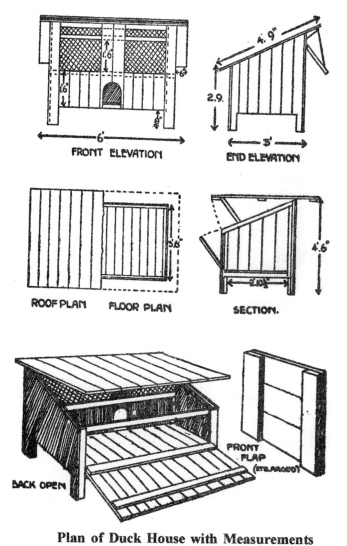

Plan of Duck House with Measurements

This should be adequate for around 5 ducks and a drake. It is designed so that the roof and back can be lifted up (hinged) to facilitate easy cleaning.

Duck Management

GENERAL PRINCIPLES

Although ducks thrive in watery conditions they must have adequate, dry, and well ventilated accommodation. Rules have been formulated which may be used as a guide, but remember that the larger breeds should have ample space so the guide below should be treated with caution.

1. Size of House

Allow at least 2 sq. feet (0.186 metres) per bird.

2. Type of House

A low lying house will suffice because ducks do not perch. However, if too low, there may be difficulty in collecting eggs and in cleaning out the shed, which must be done on a regular basis.

Good ventilation will help to keep the litter dry, but the faeces (droppings) of ducks tend to be rather wet so regular removal is essential. Soft wood shavings can be used for bedding and pebbles or sand around the entrance will help to keep the interior dry because the ducks do not carry in mud on their feet.

Keeping the interior dry also helps to keep the eggs clean because ducks tend to lay on the floor. In fact, they will lay any where -- even in a pond -- so it is advisble to keep them indoors until around 11.0 am, when they should have dropped the eggs.

Wire floors or slats have been used to allow droppings to fall through, but do not appear to be used generally; the method is not conducive to comfort because the birds are exposed to draughts from the floor.

3. Safety
The shed should be capable of being locked up each day so that foxes or other predators cannot gain access. Rats can also be a menace and for that reason the floor may be constructed of concrete.

4. Providing An Enclosed Run
If space is limited an enclosed run can be provided. This allows birds to be let out earlier because any eggs laid will be in the enclosure. A small pond or an old fashioned type of large sink or galvanized container. may also be provided to supply water.

If a hose pipe is taken into the enclosure the water containers can be rinsed out once a week so they are kept quite clean. A soak-away set in concrete will allow surplus water to be taken away.

5. Shelter From Extreme Weather
Avoid keeping ducks in an enclosure which receives the full rays of the sun in the Summer because they do not like really hot weather. Also provide hurdles or bushes to break the force of strong winds.

Duck Management

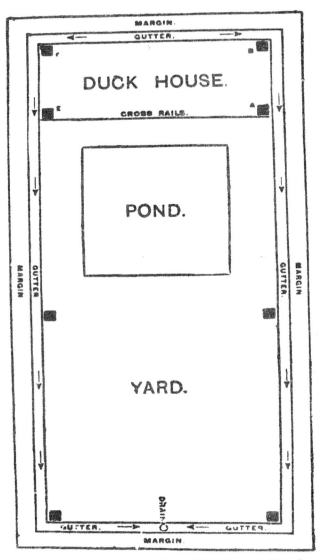

Drawing for House, Pond and Enclosure (Yard)

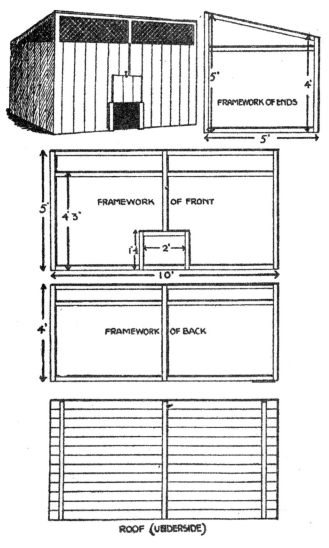

Larger Shed for around 12 Birds

Numbers to Keep

Unless special provision is made for regular cleaning and washing down or the ducks are on free range the number kept in a pen should be limited to not more than 25. This number is likely to maximize results in terms of eggs or table birds. This also means that the sheds are kept relatively small, except when producing commercially when special houses and runs are used, explained below.

The run or duck yard should be specially constructed so that the ground can be washed down with a hose. A grass or earth run is unsuitable because in a very short time the surface will be wet and covered in loose droppings.

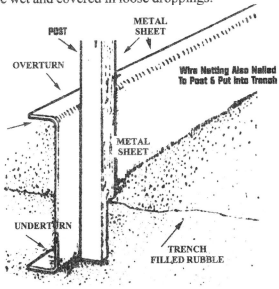

Digging a Trench to Prevent Tunnelling

Duck Management

Concrete can be used for the base, or a slatted floor through which the faeces can be washed away, or gravel to a depth of about 1 foot (30 cm) which will allow regular hose-rinsing.

The size of the run should be appropriate for the number of birds kept and generally the floor area should at least three times that of the shed. There should be a door for access and allowance made for the water container and food hopper(s).

Precautions should be taken to keep out vermin and usually the *base wire is placed below ground* and in the trench, dug to bury the wire, there should be stones or concrete. A metal sheet can also be placed at the foot of the run. Also if the ducks are to be left out at night (most unwise) the top of the high fence should be turned out so that access cannot be gained. Alternatively, the top of the run should be covered with wire netting so a fox cannot gain access.

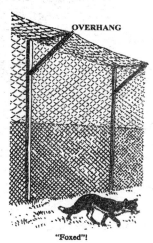

Fence Over-Hang

"Foxed"!

Duck Management 67

BROODER HOUSE FOR DUCKLINGS

Sun Parlour Shed for Rearing and Fattening Ducklings

Sheds and Runs for Intensive Rearing

COMMERCIAL DUCK RAISING

Ducks can be raised and fattened so they are ready for the table at 10 - 12 weeks of age. They can therefore be kept on an intensive basis which is not recommended for laying or breeding purposes.

The breed selected should be suitable for rapid growth and produce a plump breast in the period. Once the flight feathers are developed or when the first feathers from the neck or breast are shed they can be killed. After that period they become uneconomic, eating more food to produce feathers without significantly gaining weight. Moreover, after that period the adult plumage begins to appear and they are more difficult to pluck.

Ducklings can be reared with little or no artificial heat because they are very hardy and constantly on the move. However, the litter must be kept dry or they should be on slatted, concrete or wired floors. They should not be kept in intensive poultry houses for fresh air is just as important as ample food and water.

Breeds favoured are the Aylesbury for large ducklings or Orpingtons for the smaller size of around 2 kilos at 11 weeks old. Pekins also fatten well and are often crossed with Aylesburys for quicker growth.

For the first 24 hours no food should be given to the newly hatched ducklings. After that they can be fed on mash in a crumbly state (moisty, but not wet) five times a day. The amount is based on what they will gobble up in 15 minutes. This is done

Duck Management 69

Ideal House for Layers

This adapted duck house has excellent ventilation and a large floor area which is covered with chopped straw and bracken. The birds can be locked in at night and can stay indoors when the weather is very bad.

in the morning (7.30 am) and then 11.0 am, 2 pm, 5 pm, and 9 pm.

The number of feeds can be reduced from the second week onwards to three or four times a day. If possible they can be run on grass until 7 weeks old, but from then on they must be fed special food to add on flesh.

The accommodation can be a form of Sun Parlour or in sheds holding 25 to 50 ducklings with access to covered runs. Some form of "automatic" feeding can also be introduced, but the same principles should apply, giving the mash three times per day freshly made with warm water.

The ducklings are not allowed to swim, but they must have a plentiful supply of clean water.

Food For Intensive Rearing

The Food should have a high protein content not less than 20 per cent and Layers' Mash will supply that level.

If to be mixed from separate ingredients then the ingredients could be:

Bran (20 parts), Weatings (15), Maize Meal (4 0), Sussex Ground Oats (12), Meat or Soya Meal (10), Salt (0.50), Ground Limestone (1.50), and Cod-liver Oil (1 for first 2 weeks only).

It will be seen that Maize Meal is 40 per cent because this is quite fattening. Dried milk may also be used for mixing in the mash provided it can be purchased economically.

Duck Management

Duck Farming on a Large Scale

Water is laid on with a continuous flow in the troughs. Food is given in the 'sink' container. This could be modified. Straw is scattered in the runs.

LAYING DUCKS

The Khaki Campbells are the most popular of the laying ducks being very prolific, achieving 300 or more eggs per year. However, they are not top table birds so Orpingtons may be selected which produce around 200 eggs.

Indian Runner ducks are also excellent layers coming near to Campbells, but they do not fatten and the flesh is not of top quality.

The housing should be along the lines suggested earlier for normal production. They are best kept in small batches.

If the ducks can be let out in an orchard or paddock they will thrive well. However, they do not like disturbances or change so try to keep to a standard routine, and avoid loud noises or catching them up, which might cause alarm and affect laying.

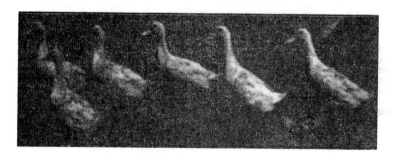

Khaki Campbells on Free Range

Duck Management

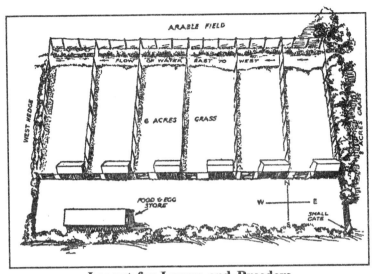

Layout for Layers and Breeders
Water from a natural stream. Simple shelters are provided.
After Reginald Appleyard

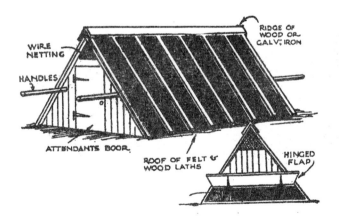

Apex Shed which is Portable

Intensive Duck Breeding
The floor is raised and slatted to facilitate easy management and the separate sections give better results.

7

FEEDING

Drawing of Utility Aylesbury Ducks
One year old; 5 kilos

PRACTICAL ADVICE

A very experienced duck-keeper who believes in natural foods gives his views as follows:

1. Give soft food in the mornings.

This should be equal parts of barley meal, bran, sharps, boiled vegetables with cod liver oil or fish meal added. This may be varied according to the ingredients available, but should be balanced so that sufficient protein is present—about 20 per cent.

Poultry layers' pellets can be fed, but these should be under cover in a suitable hopper from which the birds can scoop the food. Do not allow the pellets to become wet for a long period or they will become mouldy. Neither should they be thrown into the grass because this is wasteful.

2. Wheat should be put in a Water Trough.

This allows the ducks to take the soaked wheat when desired and does not result in waste food being taken by rats and birds.

3. Adequate Greens

Normally a plentiful supply of greens will be available outside in paddock, garden or orchard, but in winter raw swede turnips, put through a mincer will be a substitute. If they are allowed to run in an orchard, paddock or garden they will pick up insects and other tit bits and, unlike large poultry, they will not scratch a garden and cause damage.

Some ducks, including Runners and Campbells, are excellent foragers so at most times of the year can pick up worms, snails, and other tit bits, as well as greens in the form of grass and weeds.

Duck Management

FEEDING

Ducks are very hardy and in suitable surroundings will find a great deal of their own natural food such as grass, grubs and insects. However, this should be supplemented with corn soaked in water and, when laying, with layers' pellets or mash. Fattening ducklings require special feeding of high protein food.

The provision of water is vital for most breeds, although the Runners, Khaki Campbells, and Muscovies do not seem bothered and, if a pond is not supplied, there should be a water container deep enough for the ducks to immerse their heads, thus allowing them to dabble.

Grit should also be available both naturally and in a trough so that the neccessary calcium is consumed. Duck eggs are large, so a considerable supply of soluble grit is taken.

Possible Diets

There are differences of opinion on the diet to give. Some give poultry layers' pellets and corn in a water trough. Others vary the diet using different grains and additives.

Ducks lay 200 eggs or more so there should be no neglect of the food requirements. Give sufficient food to allow them to fill themselves in, say, 20 minutes and then move the troughs. Corn in water will allow them to top up in the evenings.

Sometimes the amount of food is laid down so there is in effect a controlled diet, but this is wrong because the needs vary tremendously for various reasons which are now considered.

AMOUNT OF FOOD

The amount of food required depends on a number of factors and these are as follows:

1. Body Growth & Maintenance

2. Feather Growth -- requirement heavy when moulting

3. Production of Eggs
In colder weather more food is needed.

4. Maintaining Body Heat
In winter if birds outside more food.

5. Daylight or Artificial Light Patterns
A shorter day means less food is eaten.

6. Size of Birds
As a rough guide it is usual to take 5 per cent of the body weight to calculate how much food should be given. Thus a 5 kilo bird will require 0.25 k food or 250 g per day.

FOOD CONTENT

The type of food consists of the following:

1. Protein

Generally 20 per cent protein is required for birds which are at the growing or adult stage. Newly hatched birds will require food which is higher in protein and easily digested.

Protein is expensive and if birds are fed an excess amount this is wasteful. Examples are:
- **Meat and bone meal**
- **Fish meal**
- **Soya bean meal**
- **Nut meal**

Care must be taken with meat and fish to make sure that bacteria are killed by cooking or steaming at a very high temperature.

2. Cereals

These consist of the various types of corn such as wheat, barley, maize and rice as well as meals which have been produced from these cereals. The percentage would be in the region of 65 -- 70 percent.

3. Fats and Oils

This can be provided from a variety of sources. Fish oil, rape seed oil, soya bean oil, sunflower seed oil. This will be in the region of 3 - 5 per cent. Young stock should be given extra cod-liver oil for the first two weeks.

4. Vitamins and Minerals

These are essential for all poultry to provide healthy birds. These are as follows:

1. Vitamin A
Without this vitamin growth will be retarded and certain diseases will result. Grass, maize meal and other natural products provide the requirement.

2. Vitamin D
The main provider is fish oil. Without it birds develop rickets. However, sunshine is the main source.

3. Vitamin E
This is a fat soluble vitamin which, fortunately, is to be found in the natural foods so, provided overcrowding or stress are not present there should be no difficulties in supplying the requirement.

4. Vitamin K
This is a special vitamin which prevents hemorrhages in birds. Grass or grass meal provides the requirement.

5. Vitamin B Complex
This is a composite vitamin which includes thiamin (B1), riboflavin, (B2) pyridoxine (B6), pantothenic acid, nicotinic acid, choline, vitamin B12, biotin and folic acid.

These are essential to ensure growth. Fortunately, they are provided in fish meal, yeast, distillers solubles and dried skim milk.

Duck Management

Average composition of the most common feeding stuffs

Feeding stuffs	Dry Matter	Crude Protein	Oil	Starch (Carbohydrates)	Fibre	Ash (Minerals)
Cereals, etc.						
Wheat	86	12	2	69	2	1·7
Oats and ground oats	86	10	5	58	11	3
Barley and barley meal	86	10	2	67	5	2·5
Maize and maize meal	87	10	5	69	2	1·3
Potatoes	24	2	0	20	1	—
Cereal by-products						
Wheat bran	87	13	4	53	10	6
Middlings, sharps or weatings	87	16	5	60	4	3
Oatmeal (dehusked)	91	14	9	65	2	2
Maize gluten meal	90	24	3	57	4	2·5
Animal protein sources						
Dried blood	90	85	1	2	—	2·5
Meat meal	89	72	13	—	—	4
White fish meal	87	60	4	—	—	25
Meat and bone meal	90	50	10	5	2	22
Dried skimmed milk	90	33	1	49	—	8
Dried yeast	95	49	—	35	—	10
Vegetable protein sources						
Ground nut cake meal (decorticated)	90	47	8	23	7	6
Soya bean meal	89	44	2	32	5	6
Bean meal	86	26	2	49	7	3
Pea meal	86	22	2	53	6	3
Dried green foods						
Alfalfa meal	84	14	2	29	30	8
Dried grass meal	90	23	5	40	11	10

Minerals

Calcium, phosphorous and sodium are vital. Egg production would not be possible without adequate provision. The skeleton of a bird is around 99 % calcium and phosphorous 80 % of the amounts in the body

Other requirements are manganese, salt, iodine and zinc.

All these requirements can be supplied in the rations, but for calcium and insoluble grit provision must be made to ensure an adequate supply.

Fortunately, the average poultry keeper does not have to worry too much about the complexities, provided a balanced food is made available in the correeect quantities each day.

NOTE: There are strict regulations regarding the foods that can be included by the feeding stuffs manufacturer. Bone and certain other animal offals are prohibited and the cooking must be at a high temperature. (Feeding Stuff Regulations, 1998 and amendments)

WATER

Water is supplied in the food, especially when wet mash is given. As a rough guide the water requirement will be twice the weight of the food intake. Obviously, the temperature and other factors will affect the amount taken.

There should be a constant supply of clean water before the birds and it must not be allowed to run out or to stagnate. Old,

Duck Management

dirty, stagnate water can cause severe problems. Drinking containers should be washed thoroughly and sterilized periodically. Constant running water is the ideal, controlled by the amounts taken by the birds. In the case of ducks, where practicable, a running stream can be utilized.

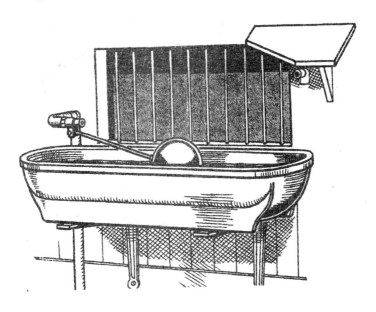

Water trough with Ballcock

The water is controlled by the rise and fall of the ballcock It must be sheltered when the weather is freezing or when there is a great deal of sunshine. The legs would be lowered to allow easy access.

LIGHTING

If production is to be maximized there should be lighting provided which is adequate for seeing to feed and drink, but not too bright. The longer day allows the stock to consume more food to produce these extra eggs. This does not mean that the lighted layers will lay more eggs during the whole year than unlighted birds. Their production will be about the same, but the lighted birds will lay heavier when under lights so that production is evened out over the year.

Birds being fattened can eat more and therefore put on flesh more quickly.

There are several systems of lighting. It can be done in the morning, or evening, or both. The system that best suits the owner's convenience can be adopted, but once a routine has been started it must be maintained throughout the winter. To alter the system or to stop it would have disastrous results on egg production.

The usual practice is to install a timing apparatus that switches the lights on and off automatically. If evening lighting is practised some sort of dimming arrangement will be needed so that, as the lights go out gradually, the birds are able to find their way back where they sleep, and are not left stranded on the floor of the house as they would be if the lights were suddenly switched off.

Suitable lighting can be provided by 40- to 60-watt electric

Duck Management

bulbs with large reflectors 3 metres apart and about 2 metres above the floor. The feeding troughs and drinkers should be well illuminated, and naturally supplies of food and water should be ready for the stock when the lighting comes on. Around 12 hours of light per day would be the aim.

The time to start the lighting is round about the beginning of October. With early hatched pullets or ducklings it could start two or three weeks before this date and would help to ward off the tendency of such stock to take an autumn moult.

With ducks the houses are often primitive shelters so proper lighting control may be difficult.

FOOD FOR DIFFERENT AGES

Ducks appear to have cast iron constitutions because they eat all types of nourishing feed greedily. However, since they do not have a beak to peck they should be fed in a manner which allows them to 'scoop' up the grain or mash. They do not have a crop or gizzard like fowls, but have a form of stomach into which the food goes for digestion.

1. Newly Hatched Ducklings

Feed them on chick or turkey crumbs dry or moistened with water.

These are fed ad lib for up to 10 days and then high protein pellets can be given followed by mash mixed into a moist, crumbly state. As noted earlier, three or four feeds per day should be given.

Water should be provided in a shallow dish, but make sure the ducklings cannot drown. Alternatively, a water fount can be used, followed by a large water fountain as the ducklings get bigger.

2. Ducklings From 2/3 weeks

Layers mash may be obtained from a mill or pet food supplier and then mixed with household scraps up to a maximum of 10 per cent. Pellets may also be obtained and kept in a food hopper for ad lib feeding. This is the quickest method and avoids complications because the ration is balance and contains all that is needed.

Duck Management

Water Fountain
Smaller size for youngsters

Food Hopper
For Pellets or Mash

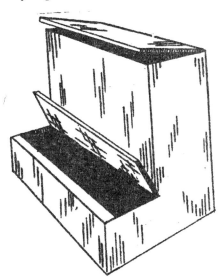

Food Or Grit Hopper

If natural foods are available with ample storing facilities then some duck breeders may prefer to mix their own foods. Examples of mixed foods is given below.

Ducks

Proved formulas for ducklings of all ages are as follows:

To be fed as all wet mash (three feeds daily)

	Breeders	Ducklings to 6 weeks	Growers (Range)	Layers (Range)	Layers (Yard or Int.)
Middlings	6	5	6	6½	5
Bran	2	2	3	2	2
Green grass meal	1¼	1¼	—	—	1½
Yellow maize meal	4	5	4¼	5	5
Fine ground oats	2	2¼	2	1½	1¼
Barley meal	1½	1	1½	2	2
Ground wheat	1¼	1½	1½	1¼	1½
Ground nut meal	¾	½	½	½	½
Fish meal	1	1¼	1	1	1
Minerals	¼	¼	¼	¼	¼
Cod liver oil (or dry supplement)	2½gls.	2½gls.	—	—	2½gls.
	20	20	20	20	20

3. Layers & Breeders

They should be fed a high protein food (around 20 per cent) with all other essential ingredients such as cereals, minerals and vitamins. For ease of feeding some use layers pellets, and corn which is put into a water-filled trough so the ducks can eat the corn at will without it being taken by crows and other birds.

Duck Management

PROVISION OF A POND

Except for achieving better fertility in some cases a pond is not essential, but water in a container in which the ducks can dabble and immerse their heads is very desirable. A typical small pond is given below for guidance of those who like to see the ducks swimming -- which is a very attractive sight and provides enjoyment when sitting watching the ducks. Remember though that eggs may be laid in the pond, especially if the ducks are let out too early in the morning.

Fibre Glass Pond

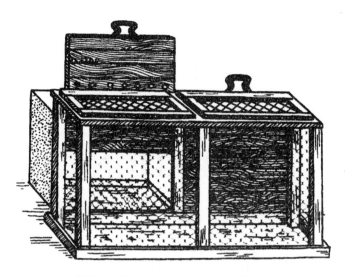

Sitting Boxes for Broody Hens

The boxes are kept in a shed and when a hen comes broody she is placed in a compartment at night with dummy eggs. This is preprepared with a turf at the bottom, shaped like a basin and a little straw formed into a nest. For duck eggs a large hen is recommended which will cover, say, 10 eggs.

8

INCUBATION & REARING

Orpington

White Wyandotte

Possible Broody Hens

Duck Management

BREEDING*

For breeding it is usual to place a small number of ducks with a drake in a run or yard. Eggs should be collected daily and kept at a cool temperature for not more than 7 to 10 days. They are then placed under the hen for the required period.

The breeding stock should be well up to standard size and not too fat. Puny stock will be quite unsuitable for producing table birds which have to be fast growing.

Breeding Pen of Buff Orpingtons
Note these are a typical shape with excellent bodies. The drake is a strong specimen. For flock breeding 25 ducks may be run with four or five drakes.

---
* Space does not permit full coverage, but see *Artificial Incubation & Rearing*, Joseph Batty, BPH

Select the Requirements

Whether breeding for showing or for utility purposes will determine the features to aim for when selecting stock for breeding. These have been designated as "Key Factors" which are generally different for each breed*. However, it should be appreciated that ducks are very much utility birds and laying abilities or fattening qualities should not be sacrificed for colour or other fancy points.

The *Key Factors* when selecting breeding birds are:

1. Laying abilities. In selecting laying breeders from Khaki Campbells, Runners, Orpingtons or one of the other breeds look for a definite bench mark or target; eg, Campbells, not less than 250 eggs and Runners similar; Orpingtons 200 eggs. This will be established from records either from a specific pen or by trap nesting. The drake should be from tested stock and line bred.

2. Size and Fattening Ability (For Table Birds).

If ducks are intended for the table they should be capable of reaching the desired weight in the period selected; eg, 10 weeks.

Select good, solid birds with no physical faults and have tight, silky feathers.

3. Colour of Egg

If white eggs are required select a breed which lays white eggs; eg, Khaki Campbell or Runner.

4. Colour of Plumage, Legs, Bill, etc.

Birds should comply with standards, but not to exclusion of utility factors.

---

* See *Poultry Shows & Showing*, **Joseph Batty, where the term was first used to indicate the main characteristics when analysing the requirements for each breed.**

Duck Management

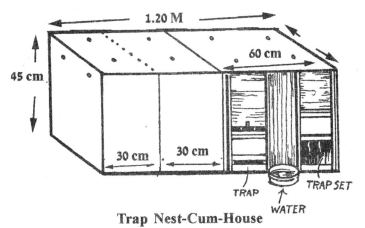

Trap Nest-Cum-House

For use in a large shed or barn. Can be used to house a duck and drake(s). Ducks lay at night or early morning so record can be made when the stock is released.

Front of Trap Nests

Both after Reginald Appleyard

MANAGING THE BROODY HEN*

The broody hen should be let off the nest once per day and then placed back on after being fed on hard corn and allowed a drink. Towards the end of the period the eggs will start to pip and the ducklings will emerge.

A steady hen is needed because the incubation period is 28 days and, since duck eggs are larger than hen eggs (80 g compared with 68 at maximum) it requires a strong hen to turn the eggs throughout the day.

Once the ducklings are hatched they should be left under the foster mother for at least a day, possibly two days, before being disturbed, although the hen may come off the nest for corn and water briefly and then go back. She may be left in the coop and run for a few days, feeding the ducklings on turkey crumbs and water on a saucer or shallow tray.

She can then be moved to a larger run so that the ducklings get more exercise and can take more food, including grass from inside the run. In fact, if this is portable it can be moved each day on grass, thus allowing the ducklings to obtain plenty of grass.

Where a number of nests are hatched together the ducklings may be grouped together and reared by infra-red lamp thus allowing the hens to go back into production.

---

* **Left to their own devices ducks will hatch their young, but broody hens are more manageable and there is no loss in production. The hen would come broody any way.**

Duck Management

ARTIFICIAL INCUBATION

The use of an incubator to hatch the eggs is now standard practice. However, for this to be worth while a large number of eggs should be available so the incubator can be filled.

Incubators are made in different sizes from 30 eggs up to thousands. The 100 egg size is probably the best size for the small producer.

RULES FOR EGGS*

1. Collect eggs daily and if dirty wash them.

If dirty they should be washed in warm water containing a mild disinfectant and when dry put them in trays with the date laid and pen number. Eggs should not be left around unwashed because bacteria develops very quickly and affects hatching.

2. Discard all Abmormal Eggs.

The wrong size, wrong shape, misshapen, and other defects the eggs should be rejected.

3. Place in Incubator Not Later than 7 days After Laid.

Fresh eggs should be placed in the machine with a cross (X) on one side to facilitate turning recognition.

4. Follow the Manufacturer's Instructions.

Place the eggs in an orderly fashion on the tray, but make sure the incubator is operating at the correct temperature before loading.

---

* See *Artificial Incubation & Rearing*, Joseph Batty, BPH for more details on incubation.

Duck Management

Special attention must be paid to humidity because ducks require extra mosture at around the 24th day. The humidity should be at 65 %. This can be checked on special instruments provided with incubators.

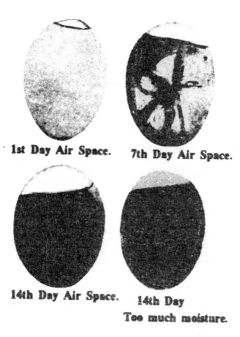

1st Day Air Space.   7th Day Air Space.

14th Day Air Space.   14th Day
Too much moisture.

Eggs with Standard Air Spaces
Photo: C Grange.

Guide for Candling Eggs
These provide an approximate guide.

Duck Management

CANDLING

At 7 and 14 days it is usual to check the eggs for fertility by subjecting them to a strong light thus seeing whether an embryo is developing. This is important because it provides a check on the fertility of the drake and allows faulty eggs to be removed.

REARING

When the ducklings are hatched place them with the broody hen in a coop or in the rearing shed with the appropriate heat. The important principle is to keep them warm, but gradually lowering the temperature until the ducklings can cope without it. In cold weather, heat may be required during the night for a short period.

Methods Available

1. Infra-red Lamps

2. Brooders which contain heated compartments

3. Hovers which are small boxes in which a lamp or heating element is placed.

The first three days are critical because the ducklings must be kept at a fairly high temperature. At the beginning the temperature should be 35^0 C and this should be adjusted each week so that the temperature reaches around 21 degrees in about 4/5 weeks. A heated compartment can be maintained up to that point but a cool section adjoining should contain the food and water. In this way the ducklings become quite hardy and independent.

In the house there should be bedding in the form of shavings or chopped straw, thus absorbing the moisture. When hardened they can be allowed outside on to grass, making sure they are well protected from predators. A Sussex Ark and portable run would be ideal.

Infra-Red Lamp Brooding

Duck Management

Rearing in a Large Shed

From about 4 weeks, unless very cold weather, the ducklings can be reared without heat.

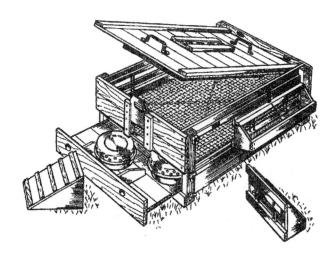

Brooder for Rearing
Lamps or Electricity would be used

9

MANAGEMENT POINTERS

Duck Management

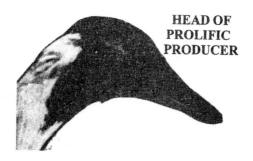

Head of Healthy Duck

Signs of Condition on the Water

SOUND MANAGEMENT

If excellent results are to be obtained then sound management is essential. This covers selection of breed of the correct type, suitable accommodation, proper feeding, and observation of the ducks on a daily basis. Any falling off in egg production, food being left, birds appearing out of condition, specific ailments* or other out of the normal factor should call for investigation.

SELECTION OF GOOD PRODUCERS

The breeding stock should be of top quality without faults because generally 'like produces like' and any weaknesses in the parent stock will show up in the young.

Birds should feel solid, but not excessively fat. The feathers should be smooth and glossy and close to the body.

The breast should be rounded without an obvious keel and on the layers the abdomen should be full and fairly low down, but not to the extent of being a prolapse.

The head of the producer should be tightly feathered, with bold, bright eyes, set high up on the head.

Observing the ducks on the water can provide clues on the condition of laying stock. The laying duck tends to swim with her tail *not* raised; she will also have a well developed abdomen

On the other hand, the duck which is not laying will look rather smooth in outline and the abdomen region will not be fully developed, which is proper for a young duck under 6 months old, but not a mature layer.

---
* See *Poultry Ailments* ..., J Batty, BPH

AVOID OVERCROWDING

Ducks do not like to be too crowded and therefore they are best in small units up to 25 and certainly never above 50. Neither do they take to fully intensive conditions such as those used for poultry which are often kept in battery cages.

Leave sufficient space in the food troughs and water containers to allow all the ducks in a pen to take food and water without being pushed to one side.

If a large number of birds are being kept then use a very large bucket-type container made of metal which will allow the food to be mixed. Indeed, if very large quantities are needed a small cement-mixer type machine can be used. If mixed by hand a small, well-worn shovel will be needed and normal sized buckets for taking the mash to the birds; a wheel barrow or trolley can be adapted for this purpose.

There should be a mixing room in which the dry mash is kept in metal storage bins. Any other ingredients should also be close by, as well as a large, deep sink with a flat bottom so that water can be obtained and utensils can be washed.

Eggs can also be kept nearby on trays, separating the eating eggs and those for hatching. Make sure the eggs are clean and that they are used up within 10 days.

If on a large scale an egg washing maxchine can be used which allows large numbers to be washed.However, do not get carried away by trying to wash every egg because natural cleanliness is better.

Duck Management

Room for Storing Eggs and Keeping Records

In another part there can be a mixing area, but partitioned in a separate part.

Egg Washing Machine (Rotomaid)

FREE RANGE

If ducks are kept in large numbers they should preferably be on free range where they can obtain a great deal of natural food. This is ideal for layers for producing hatching eggs or for the market.

Another method, covered earlier, is to have specially built pens so the ducks are kept on a semi-intensive basis, but still allowing them outside.

Ducks will thrive on all kinds of grass land, but this should not be in a constantly muddy state because this is unhygienic and would not allow grass and other foods to be taken.

Many ducks will be fertile on free range, but, if there is difficulty, a small pond can be provided.

Ducks on a Pond -- improves fertility

Duck Management   109

Aylesbury Ducks feeding on Grass Land

The general management consists of having distinct pens for each group of birds, based on the number hatched at a particular time, so that they are all of the same age, and feeding can be controlled better.

HEAVY BREEDS FOR FATTENING

Certain breeds are more suitable than others for rearing for table birds. There will also be strains for each one which give excellent results, usually because they have been bred and developed for the purpose of producing fast growing birds.

The breed which suits the market should be selected. In terms of relative merit the Aylesbury is usually the best because some strains will be ready for the table at 8 weeks (2.25 k).

Pekins and Orpingtons give good results in about 10/12 weeks. The Rouen provides a better flavoured flesh, but may take 7 months for the frame to fill out so it is better for situations where a larger carcass is required.

The Muscovy is an excellent table bird, but does not mature at a fast rate. The flesh is dark and strong in flavour.

Watching For Fertility

The table birds should not be coarse and heavy boned with large heads and beetling brows. Such birds are not likely to be active foragers and therefore not likely to be profitable.

If not active the drakes may not be fertile. They certainly should be from March or a little earlier, but without the necessary vigour they are likely to fail.

Drakes can be left with the ducks all year round and this does not seem to have adverse effects on results. Providing light in sheds can stimulate drakes as well as the ducks.

Duck Management

Ducks for the Table

1. **Aylesbury**—best table breed.
2. **Pekin**
3. *(Bottom right)* **Rouen:** excellent but slow to mature.
4. **White Pennine**; small, plump duck, but now quite rare.
5. **Cayuga**
6. **Muscovy**; tends to be rather strongly flavoured, but a good size. Is a separate race and incubation is longer -- usually 35 days.
7. **Orpington**

RINGING AND MARKING

For birds to be recognized for trap nesting or keeping records some form of marking will be necessary. The most usual methods are:

1. Toe Markings when small V-shaped nicks are made to provide information on each duck. Toe punching may also be practised.

2. Ringing which can be made of plastic or aluminium.
If the closed type they are placed on at about 6 weeks. Plastic split rings can be placed on at any time.

3. Wing Bands which are fixed to the wing and contain information such as date of birth and other details.

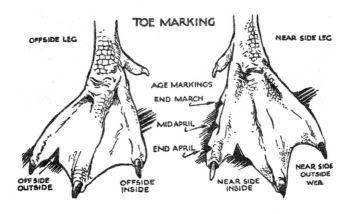

Toe Markings -- may be adapted to own system

Duck Management

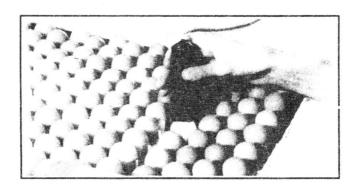

BRISTOL EGG CANDLER (Hand Held)
Easy to use and the eggs do not have to be handled.

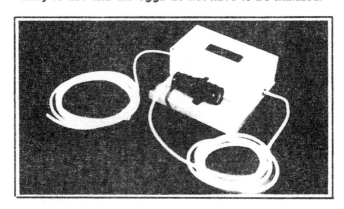

Egg Candling Machine
Complete Tray can be candled so eggs do not need to be handled individually.
(Courtesy: Patrick Pinker Game Farms, Bristol)

KILLING BIRDS

Birds should be killed and plucked and it is usual to restrict food, but not water, for 24 hours before killing takes place.

The killing should be done as humanely as possible and no processing should take place before birds are dead. Where the birds are sent alive to a processing plant they are placed in a cone or hooked on to a moving overhead chain and the killing and processing follow in quick succession.

For the small producer the killing is done by one of the following:

1. Dislocation of the Neck

The duck is held by the legs in one hand and the head in the other and a steady downward pull results in the dislocation and death. Just the right amount of pressure should be exercised because if the pull is too strenuous the head will be torn off.

2. Cutting Throat and Allowing Bleeding

The jugular vein is cut and blood pours out leaving the flesh white. However, there must be provision for catching the blood.

Take care when catching the birds because any panic or rough handling can damage the carcass. A catching crate can be used to drive the ducks in, when they can easily be removed.

Duck Management 115

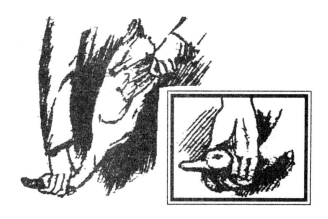

Killing A Duck by Dislocation
Note position of hands on legs and neck

PLUCKING

Take care when plucking which can be by hand, by machine or by using wax. Each method has its merits and, obviously, for any large scale operation a dry or wet plucking machine will be needed.

Once killed a duck should be placed on a board with its breast facing down so it is shaped properly. In a factory this would be done when each duck is placed in a plastic freezer bag which is printed with the details, including weight.

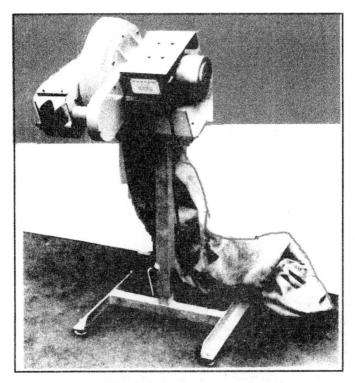

Dry Plucking Machine (Cope)
There is also a wet plucking machine and a machine which uses hot wax.

INDEX

INDEX

Accommodation 59 - 75

Aylesbury 68

Breed (Detailed) 1-48

Breeding 93-, 110

Breeds 49, 50

Broodies 90, 92, 96

Candling 98, 99, 113

Classification 51-56

Commercial Duck Raising 68, 73

Cross Breeding 57

Diets 77 (See Feeding)

Eggs 97, 106

Egg Washing 106

Fencing 66

Feeding 68, 70, 75 - 90, 106

Food Content 78-, 81

Food Different Ages 86
Food Quantities 78, 88
Free range 108, 109

Incubation 91-102
Indian Runner Ducks 58
Intensive Management 67, 71

Khaki Campbells 72
Killing Birds 114

Lighting 84
Layers 69, 72

Management 117
Minerals 82

Orpingtons 68

Plucking 115, 116
Ponds 89

Rearing 91-102, 99

Ringing 112

Runs 66

Space 106
Standards 1-48
Sun Parlour 70

Table Birds 110, 111
Toe Marking 112

Trap Nests 95

Utensils 87

Vitamins 80

Water 82-

Wing Bands 112